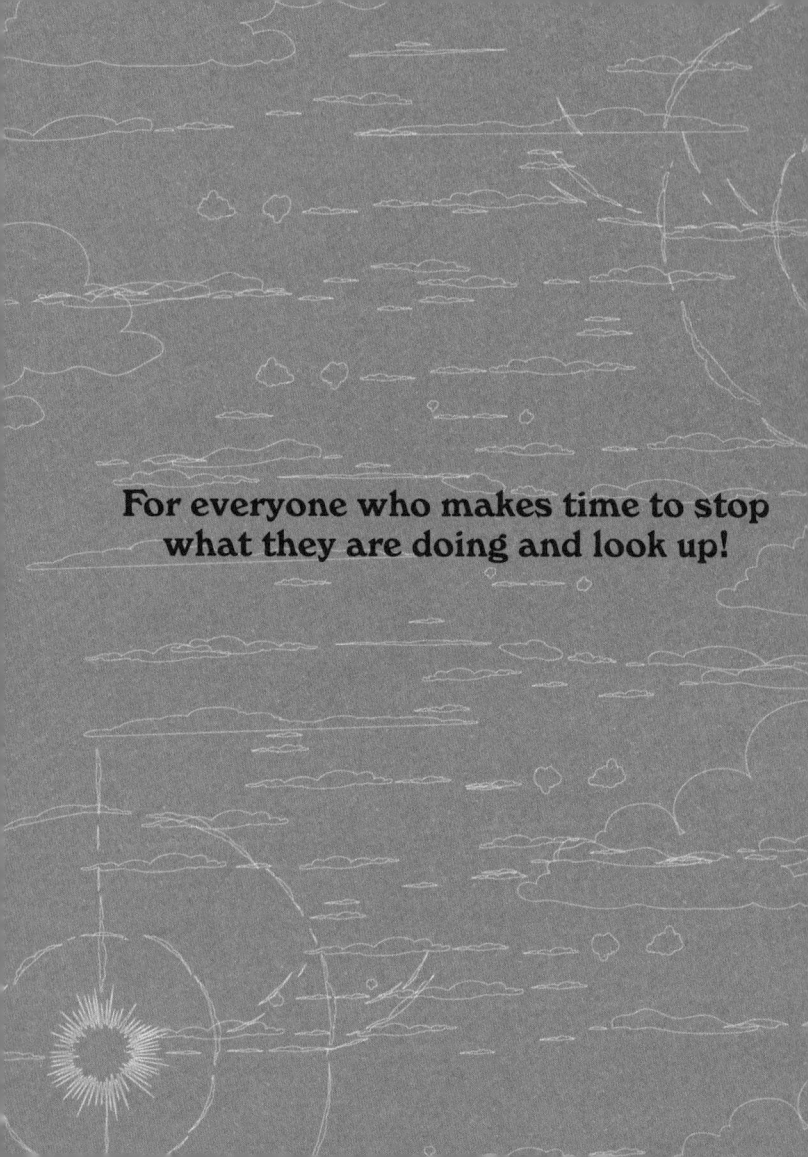

For everyone who makes time to stop
what they are doing and look up!

CLOUDS

A GUIDE FOR
THE CURIOUS

Susan E. Clark

Quadrille

PART ONE
ALL ABOUT CLOUDS

PART TWO
CLOUDS
TO SEE

When we think about the wellness benefits – physical and mental – of getting outdoors, we tend first to think of all things green and leafy. We might then remember the glorious palette of the vast oceanscape that covers our planet, and the rivers and waterways closer to home that teem with hidden lifeforms. Rarely will we think to look up. This is a shame, because cloudspotting, or even cloud-collecting (keeping a journal/record of those you have spotted), can bring all the benefits of a satisfyingly deep Yoga Nidra relaxation – only outdoors.

Clouds have captured the imaginations of weather scientists, poets and painters alike since the dawn of time. There's something about the transience of a cloud pattern that reminds us that in life, everything is about flow, and nothing stays the same.

One minute, we are looking at the beautiful puffy, fluffy, cotton-like and flat-bottomed Cumulus clouds (page 52) and the next, these have risen higher in the sky and given way to Stratocumulus (page 60), which is the most common cloud in the world.

And so, this beautiful guide for the curious will help you learn more about some of those stories. The word curious comes from the Latin *cura*, which means 'to care', and once you start getting curious and really looking at the world around you and connecting more deeply with the natural world, you will soon find you really do care more about what's happening to the world outside your door and the wider planet.

Be warned though, cloudspotting is completely addictive, because once you look up to the heavens and begin to learn to 'read' cloud patterns and signs, you won't be able to stop. And as you learn more, you'll find yourself becoming quite the amateur weather forecaster – which means you'll always know what to wear for the day ahead!

Disclaimer:
This is not an identification guide, but a book to make you look up and get more curious about what's overhead.

PART ONE
ALL ABOUT CLOUDS

WHAT IS A CLOUD?

Clouds appear when there is too much water vapour (gas) for the air to hold. We are surrounded by water vapour all the time and although we cannot see it, we can feel it. When there is more in the air, it feels humid and muggy and when there is less, it feels dry and fresh and just right for going outdoors. But when there is too much water vapour, the gas condenses and forms tiny water droplets that are visible. But they are so tiny – about one-hundredth of a millimetre in diameter – they stay suspended in the sky.

These droplets, which cling to little bits of dust in the air, are what we are looking at when we look up and see clouds.

WHERE DOES A CLOUD COME FROM?

Imagine you are trekking up a mountain or a very big hill. Notice how, as you ascend, the air around you starts to feel much cooler. Colder air cannot hold as much water vapour as warmer air and so as the air cools, it becomes saturated and the water vapour it holds condenses. This simply means it turns from a gas to a liquid – something you will have seen for yourself when you notice the wet condensation trickling down a cold glass window.

If the cloud is high enough in the sky and the air is cold enough, these droplets turn to ice crystals and then instead of seeing plump, fluffy clouds, you will see wispy trails and tendrils streaking overhead.

Clouds will also form in places where they will likely pick up more water vapour such as over a lake or other large body of water.

WHAT'S IN A NAME?

Until the 1800s, there was no agreed way – other than using poetic terms like 'wispy' or 'menacing' – of describing what the eye could see when we looked up at the clouds.

But all this changed when, in 1802, a pharmacist and amateur meteorologist named Luke Howard (1772–1864), whose real passion was cloudspotting, suggested scientific names for the then three main types of clouds. This paved the way for the scientific classifications we have today.

THE CIRRUS	THE CUMULUS	THE STRATUS
for curl/ lock of hair	meaning heap	for layer

He then went on to add the following sub-categories:

CIRRO-CUMULUS	CUMULO-STRATUS
made up of small cloudlets sprinkled across the sky	a lower layer of patchy cloud with a well-defined base

CIRRO-STRATUS	CUMULO-CIRRO-STRATUS
made up of a layer of ice crystals that give a whiteish hue to a blue sky	which is also called the nimbus or rain cloud – made up of water droplets that turn to ice crystals to give us a dark and brooding grey sky

Howard, who became famous for naming clouds, was a good watercolourist and would illustrate the different cloud types he observed on his regular travels between London and the UK's Lake District. He was so admired for his commitment to developing a proper system to classify clouds, the great poet Johann Wolfgang von Goethe (1749–1832) even wrote several poems dedicated to this new system of classification including one which he wrote in 1815 and called *In Honour of Howard* and in which he paid poetic homage to the big four main cloud types: Stratus, Cirrus, Cumulus and the later addition, Nimbus.

ALTITUDE AND ATMOSPHERE

The atmosphere is a mixture of gases that surround Earth, without which there would be no life on the planet. These gases are held in place by gravity.

There are seven different layers of the atmosphere, which extends up from the Earth's surface for around 10,000 km (6,214 miles).

The layers where most clouds form are:

THE TROPOSPHERE

This is the lowest layer and the one we live in. It is the most humid of the layers, so this is also where most clouds form, usually between 8 and 14 km (5–9 miles).

THE STRATOSPHERE

This layer extends from the troposphere up another 50 km (31 miles). The air is very dry and contains little moisture, so only a few clouds can form.

The remaining cloudless layers are the mesosphere, the thermosphere, the ionosphere, the exosphere and the magnetosphere. Noctilucent clouds are the highest of all the clouds, forming in the mesosphere at about 85 km (53 miles). These thin clouds are composed entirely of ice.

Clouds are also classified as high, mid or low level. The latter are usually 2 km (1.25 miles) above ground level, mid-level clouds form between 2 km (1.25 miles) and 4.5 km (around 3 miles) and the high-level clouds form above that. Cirrus (page 50) are the highest forming clouds you're likely to see, although the show of spectacular lights known as the Aurora originates even higher up in the thermosphere.

THE TEN MAIN CLOUD TYPES

There are ten main cloud types to remember. You'll find a detailed description of each in Part Two (pages 48–109) but here, briefly, is how they all affect the weather.

1. CIRRUS

High, wispy clouds that let
most of the sunlight through.

2. CUMULUS

Sharp-edged,
cotton-wool-ball clouds that
block most of the sunlight.

3. CIRROSTRATUS

Thin, transparent clouds
that create a milky blue
sky and allow
sunshine through.

4. NIMBOSTRATUS

Thick grey rain clouds that
completely blot out the sun.

5. CIRROCUMULUS

Gently rippled clouds that allow
sun through but usually precede
stormy weather.

6. STRATOCUMULUS

Low clumpy and patchy layers
of grey and white clouds.

7. STRATUS

Low, thick, grey
blanketing clouds,
that block the sunlight.

8. ALTOCUMULUS

The sun (or moon) will be vaguely
visible through these clouds.

9. ALTOSTRATUS

Boring-looking grey sky so not
much to get excited about.

10. CUMULONIMBUS

You won't see the sun or the moon or
very much else once these dark and
thunderous-looking extreme-weather
clouds take over the skies.

SPECIES AND VARIATIONS

There are clouds that wave and clouds that wisp and even clouds that create great castle-like turrets in the skies.

That's because some of the ten main cloud types can come not only in a range of different species but also in a range of different variations.

We'll explore some of these exciting varieties in Part Two, but for quick reference: cirrus, stratocumulus and altocumulus all have a total of five associated species, whereas altostratus and nimbostratus have none. And the truth is, the same cloud type or the same cloud type species can show up in what can seem like an infinite number of varieties, making it impossible to name them all.

Don't worry if you're feeling overwhelmed and confused just thinking about these classifications. For us meteorological amateurs, it's enough just to know the ten main cloud types and, for bonus points, a handful of the more obvious variations and optical effects, all of which can be spectacular to see and fun to know more about.

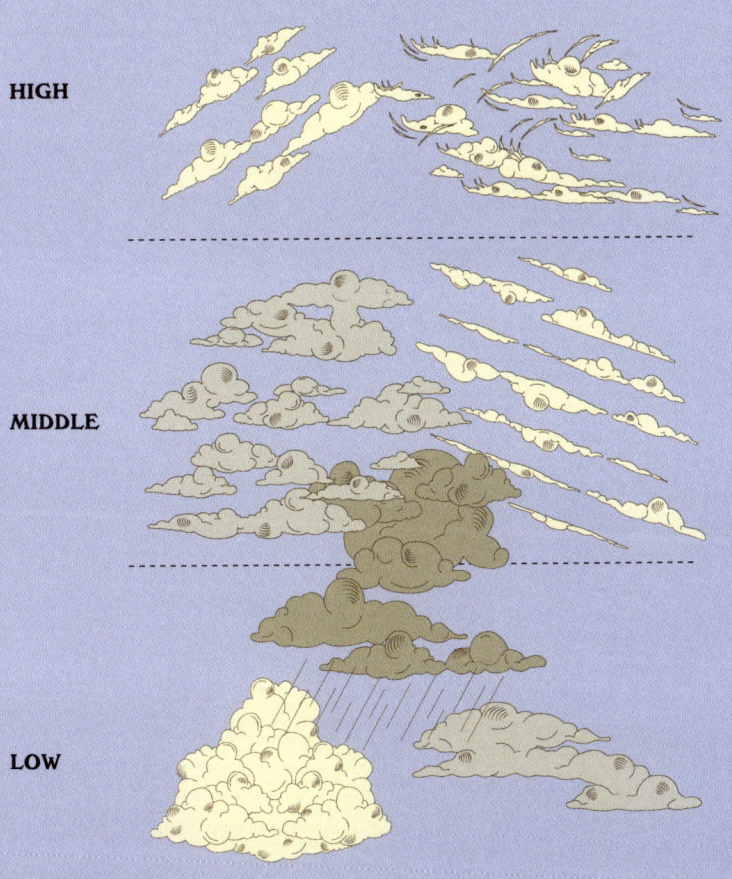

HIGH

MIDDLE

LOW

RAIN

It rains when air rises into the upper atmosphere (page 11) and cools. The cooler temperature at this level causes the water vapour the air is carrying to condense into water droplets and then, once the air becomes saturated, these fall from the clouds as rain. Air pressure, as in the weight of the air in the atmosphere, contributes to this process.

The water in air comes via evaporation from the Earth's oceans, lakes, forests, fields, plants and animals, and so rain is the way this life-giving resource is returned to replenish our water sources. It also provides water for plants and animals – including us.

Can you imagine how things would look and how life would be if it didn't rain where you live for a whole year?

Rivers would dry up and plants and animals would perish and with continued drought extreme desert conditions would develop across large swathes of the planet; all the greenery we take for granted dried and died and disappeared. Crops would fail and your grocery shelves would show how the world had plunged into panic and lack.

So, unless flooding is a threat to life where you live, put on your raincoat and boots when it rains and do a happy dance to say thank you to the clouds and the natural world.

THE 'SPUN UP' WATER CYCLE

Rising global temperatures will likely accelerate evaporation worldwide, which means – if you take a global average – we are likely to see more rain in coming years. Scientists call this a 'spun up' water cycle – a direct result of global warming.

Of course, there is no guarantee this increase in rainfall will be distributed evenly across the planet, so what we will see is more flooding in some parts and more drought in others – a shifting baseline showing movement in what were the traditional rainbelts and desert regions.

The truth is, we don't yet know for certain. Some modelling shows increased rainfall over oceans but not land, others say coastal regions will become wetter while inland areas will become drier.

Another factor to consider is that as temperatures warm and levels of carbon dioxide increase, we may see more water released into the air via transpiration, which is the release of moisture by plants as a result of photosynthesis. More plants, and more plants growing more rapidly, will result in an increase in the rate of water uptake from the soil, resulting again in drier ground in that region.

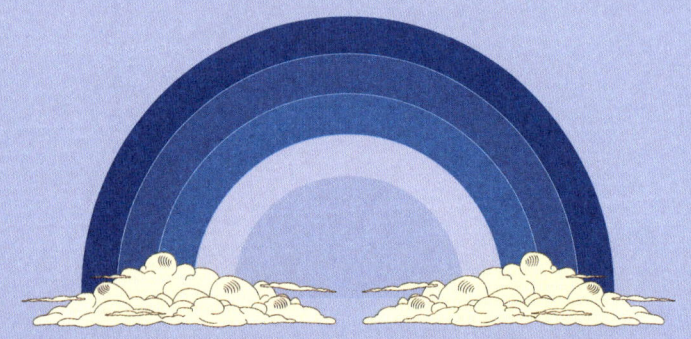

RAINBOWS

As small children, we don't need to be told that rainbows are magical, and we may even set off on an adventure to seek that mythical pot of stolen gold said to be buried at the end of every one.

But did you know there's more than one type of rainbow?

A rainbow is an optical cloud effect that forms when direct sunlight shines from behind you onto a rain shower up ahead. The light that passes through the falling raindrops reflects off their back surface, separating into those seven well-known rainbow colours as it passes in and out of the water.

You may have seen a double (or secondary) rainbow, which happens when the light bounces off not one but two places at the back of the raindrop. Or you may have seen a low rainbow, which appears when the sun behind you is high in the sky and you can only see the top of the bow above the ground.

When the raindrops themselves are small, you may see what's called a supernumerary rainbow, where there are multiple tiny lines appearing on the inside of the main rainbow arc. Or perhaps you've seen a strangely zig-zag-shaped reflection bow, which happens when sunlight reflects upwards from still water onto the rainbows.

One of the more mystical bow effects is the ghostly white rainbow. This ethereal spectre appears when the light hits tiny droplets in fog.

SNOW

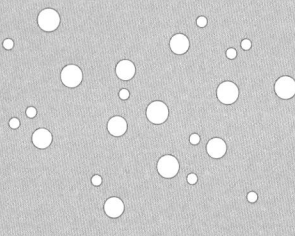

As long as there's no threat to life, snow can be one of the most magical of our weather effects, covering the ground with a pure white blanket (until it becomes a murky slush!) and bringing a kind of welcome hush to the world outside your front door.

If you're a kid or even an adult 'kid', you'll be itching to get the sleigh from the garage and if it happens to snow at Christmas, well then everyone will be whistling Bing Crosby's famous song and feeling that special seasonal goodwill.

Snow is a form of precipitation made up of beautiful ice crystals that form when water vapour in the air freezes. These then grow while suspended in the atmosphere, before falling to the ground. In short, we need two conditions for snow: temperatures below freezing and enough humidity in the air.

As well as gracing our Christmas cards and inviting us outdoors to play, snow has an important function in controlling patterns of heating and cooling over the Earth's surface. Ground that is not covered with snow will absorb up to six times more of the sun's energy and will therefore be warmer than ground that is blanketed by snow.

Snow is also an important source of drinking water and hydropower in some regions. And of course, it can be deadly. Compacted snow debris can weigh more than 500 kg (1,102 lb) per cubic metre, which is half a ton. Those unlucky enough to be buried by an avalanche have a 90 per cent chance of survival, but only if found in the first 15 minutes.

And if you've ever wondered why some clouds are white, it's because their billions of water droplets or ice crystals reflect the light in all directions and our eyes perceive this as the colour white. Snow looks white for the exact same reason.

ICE-CRYSTAL SHAPES

Just like snowflakes, no two ice crystals look the same
– although they all have six sides.

In air temperatures between -10 and -40°C
(14 and -40°F) a cloud will be a mix of water droplets
and ice crystals.

Below -40°C (-40°F), it will be made entirely from
these beautiful ice crystals.

HAIL

Hail is supercooled water that is refrozen in thunderclouds in the atmosphere before falling to the ground as a sizeable ice ball.

Hailstones are usually spherical or conical. Most will range in diameter from 5–50 mm (½–2 in) or even more, but the majority are smaller than 25 mm (1 in). The largest hailstones tend to form when shower clouds cluster together to form 'multicell' storms, which are common in the United States.

How fast hail falls depends on the size of the hailstones and the surrounding air and wind speed, which means it can fall from anywhere between 15–160 kmph (9–100 mph).

The world's heaviest hailstone was recorded in 1986 and weighed in at 1.25 kg (2¼ lb). Imagine a large grapefruit falling from the sky and you'll get the idea of how much damage or injury this could cause. This particular hailstorm happened in Gopalganj in Bangladesh and killed 92 residents.

FOG

If you're a director looking to add menace and drama to a movie, then fog is your best friend – just ask Alfred Hitchcock, who often used fog to heighten anxiety and threat and made at least two films – *The Lodger*, a 1927 silent thriller, and *Fog Closing In* (1956) – where you might argue that fog was the real star.

Sadly, for most of us, fog is not our friend, but rather a day to stay indoors and wait for the gloomy skies to pass.

The thickest fogs happen in areas of high pollution, over cities and industrial areas, because the pollutant particles hanging about in the air allow the water droplets to coalesce and thus grow in size. A ghoulish arc of fog is known as a ghost rainbow (page 25).

If fog resembles any of our main cloud types it is gloomy, grey Stratus (page 62).

CONTRAILS

Contrails – short for condensation trails – are long bands of ice-crystal trails that criss-cross the skies behind an aircraft cruising at high altitudes. They are created by the aircraft's engine fumes, made from water vapour that becomes ice crystals and appear typically 7,500 to-12,000 m (25,000–45,000 ft).

Sometimes, a contrail will disappear almost as soon as you catch sight of it, other times it may remain visible for several minutes.

The eagle-eyed may notice there is usually a gap between the tail of the aircraft and the start of the contrail. This gap of 50–100m (165–330 ft) is because the exhaust coming from the engine of a jet aircraft is warm and moist, and the water vapour content comes mostly from the combustion of hydrogen in the aircraft's fuel. As they hit the atmosphere these gases cool rapidly – within a fraction of a second – and it is this slight delay that results in the gap between the craft and the contrail.

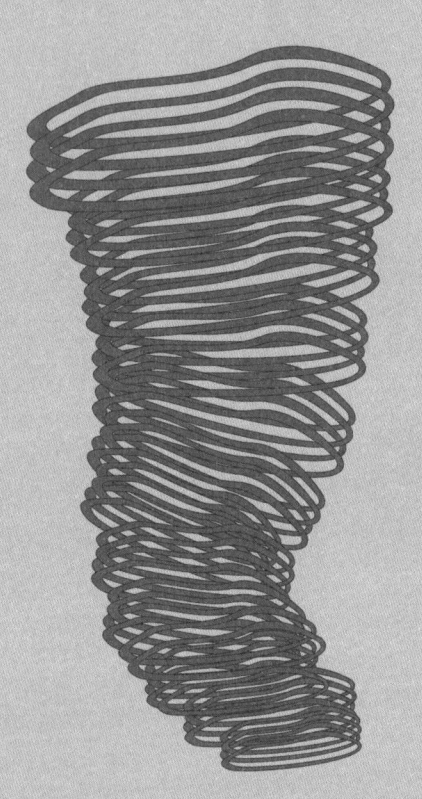

TORNADOES AND TWISTERS

The 2024 action film *Twisters* is a sequel to the 1996 film *Twister*. What this tells us is that it's not just Hollywood who has a fascination with these life-threatening weather patterns, but cinema-goers too!

A twister or tornado is a violently rotating funnel or tuba cloud (page 100) that reaches the ground and is also in contact with a stormy cumulonimbus cloud (page 68) up above. Debris, dust and anything in the path of the bottom of the funnel gets picked up and tossed about, which is what makes a tornado so dangerous and to storm chasers, irresistibly fascinating.

The winds of a tornado can reach up to 480 kmph (300 mph) – that's strong enough to rip the roofs off houses, uproot trees and toss a car into the air. Most tornadoes last for less than 10 minutes and travel around 5–10 km (3–6 miles), but some may last several hours and cover 150 km (93 miles)!

Tornadoes are measured using the Fujita or F-scale which ranges from 0–5, with five being the strongest. The deadliest tornado on record to date happened in Bangladesh in 1989. It tore through twenty villages in the Dhaka region and killed more than 1,000 people.

ACCESSORY CLOUDS

Sometimes our main cloud types show up accompanied by other, usually smaller, clouds, which may appear unattached to the main cloud or partially merged. These are what we call accessory clouds.

An accessory cloud is always dependent on the main cloud type for its development and continuation and so will never appear alone.

Pileus or cap clouds (page 86) are an excellent example of an accessory cloud that, in this instance, sits over the top of a main cloud.

Within the ten main cloud genera there are fifteen cloud species, nine varieties, eleven supplementary features and four accessory clouds.

**ACCORDING TO THE
INTERNATIONAL CLOUD ATLAS,
THE THREE OF THE FOUR ACCESSORY CLOUDS
(ALL OF WHICH FEATURE IN THE BOOK) ARE:**

**Pileus (page 86)
Velum (page 88)
Pannus (page 92)**

SUPPLEMENTARY CLOUDS INCLUDE

**Mammatus (page 73)
Praecipitiato (page 94)
Virga (page 96)**

Asperitas is a rare ripple-shaped cloud that has the distinction of being the first new cloud added to the *International Cloud Atlas* in fifty years when it was formally listed as a supplementary cloud in March 2017. This was thanks to the campaigning efforts of the Cloud Appreciation Society and its founder, Gavin Pretor-Pinney.

CHANGING CLOUDS, CHANGING CLIMATE

Scientists report that as the climate warms, the amount of each cloud type is also changing and, according to scientific modelling, it is thought clouds may even amplify climate change in the future.

The researchers describe a kind of loop whereby a warming climate will cause more clouds that trap heat, which in turn will create more warming. This is known as positive feedback, although you can see that in this instance, positive does not mean good. In fact, it means the problem of climate warming is compounded. A better term to use would be 'vicious cycle', which everyone understands is not good. Remember too that while some clouds such as cirrus (page 50) do have a warming effect on the Earth, others such as stratus (page 62) will help cool it.

WHY DO WE LOVE CLOUDS?

Nephophile:
a person who loves clouds.

Clouds can reflect and even influence our emotions. How do you feel when you look up and see a clear blue sky on a warm sunny day? How is that different from looking up and seeing thick grey clouds gathering on the horizon and heading your way?

As well as being beautiful and often enchanting and magical, clouds drifting across the sky can bring a sense of calm and oneness with the natural world, which is good for our wellness and mental health. Plus, clouds are just gorgeous.

When there are interesting clouds about, head out to the highest hilltop with a picnic and a blanket to lie on so you don't crick your neck. Spend a few hours noticing the calming effects of cloud gazing.

CLOUDS IN ART

The English landscape painter John Constable (1776–1837) loved painting clouds and produced over 100 paintings of his observations of cloud patterns, which he even made up a name for: *skying*.

Constable was a painter in what was called the Romantic tradition; this is hardly surprising since there is something so inherently romantic about the drifting of clouds across the sky.

Taking care to note down the precise location and weather conditions on the back of each of his cloud sketches, the painter elevated the status of clouds when he wrote, 'We have had noble clouds and the effects of light and dark and colour.'

In other words, everything a landscape painter could ask for!

CLOUDS IN MYTHOLOGY

In Greek mythology, the Nephelai were the nymphs of clouds and rain. They arose from the Earth-encircling river Oceanus, bearing water from the heavens (rain) in pitchers made from clouds, to feed the streams of their River-god brothers and thus nourish the earth.

Their ruler was Zeus, King of the Gods and ruler of the heavens. He was the god of clouds, rain, thunder and lightning.

CLOUDS IN THE SKY – A SIMPLE CONNECTION PRACTICE

The drifting by of clouds – whatever their type – reminds us that nothing lasts forever, nothing stays the same and nothing really belongs to us. And if you can accept and integrate this fundamental truth, you will find a happiness and deep inner peace, thanks to what Buddhist practitioners call non-attachment.

Clouds show us how to let go, let alone and let be. They show us that as ever, nature has much to teach us if we can just get curious enough (and care enough) to listen and to learn.

'Clouds in the sky', as a simple and mindful connection practice, quite simply is remembering to take a few minutes of each day to stop what you are doing, look up over your head or out of the window and notice the clouds that may be hanging about or drifting by.

Remember you are as much a part of their world as they are of yours. Notice how your breath and nervous system respond when you remember this connection and then go about your day knowing you are as much a part of – and one of – the wonders of the natural world as the clouds above you.

PART TWO
CLOUDS
TO SEE

Angelic-looking and ethereal

CIRRUS CLOUDS

If angels had art classes and the subject was clouds, they would probably paint cirrus, which are the most heavenly-looking of all the cloud types.

Falling down through the high winds of the lower layer of the Earth's atmosphere (known as the troposphere), they trail delicately across our skies and foretell, as they gather, the welcome and imminent arrival of a warm front.

Made almost entirely from plate-shaped ice crystals, to the eye these clouds look like the flowing white locks of the hair of an elderly giant, and in fact, their name is a Latin word which means 'a lock of hair'.

And if you still need convincing of the celestial connections, sometimes cirrus clouds will refract and reflect the sunlight to produce coloured arcs and rings which cloud collectors call halos.

Cirrus clouds form when warm, dry air rises, causing water vapour to deposit onto rocky or metallic dust particles at high altitudes.

They are composed almost entirely of ice crystals.

Seeing cirrus clouds thickening overhead may be a sign of rain or snow on its way.

Optical phenomena when sunlight interacts with ice crystals, creating effects including 'sun dogs'. These appear with a wind chill between -20 and -40°C (-4 and -40°F).

Clouds that drift by

CUMULUS CLOUDS

You know those cotton-wool clouds that drift lazily across a blue sky on a sunny-ish day? Those are the cumulus clouds, which are easy to recognise because with their lumpy, bumpy tops they look like the creamy, white head of a cauliflower.

And when you do spot them, notice how the base of the cloud is darker and how those cauliflower tops are a brilliant white when the sun shines through them.

You can think of these as mostly fair-weather clouds – they don't foretell rain or snow is on its way, unless the cauliflower tops begin to build upwards in denser-looking towers, which means the weather is about to change. When this happens, these clouds are known as cumulus congestus.

Cumulus congestus can produce those intense but short-lived rain showers that catch you out. And if these clouds keep growing into cumulonimbus (page 68), you can expect stormy weather.

Cumulus clouds change shape even as we gaze at them, so unleash your imagination and see what shapes, creatures and faces you can see in them as you watch them drift by.

Cumulus is a Latin word that means 'pile' or 'heap'.

These puffy clouds form on thermals – those invisible columns of air that rise from the ground as it is warmed by the sun.

Cumulus clouds form a few hours after dawn and are usually gone before the sun goes down on another day.

The higher the grey base of each cloud hangs in the sky, the drier the atmosphere and the fairer the weather will be.

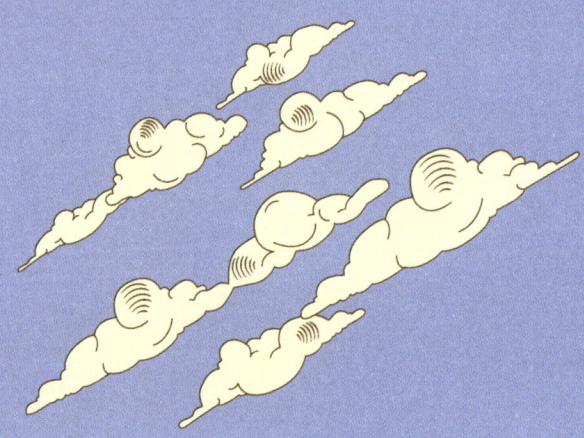

Understated and therefore easy to miss

CIRROSTRATUS CLOUDS

This gorgeous cloud is one of the more subtle of the ten main cloud types and so – because it doesn't really look how we imagine clouds to look (unlike cumulus, page 52) – we may not even notice it hanging high in the sky.

Made from a delicate layer of ice crystals spread over a vast expanse of sky, what you will see when you look up is a milky, whitish lightening of the blue sky and, if you are lucky, an impressive halo. These are the clouds that can produce a range of arcs and rings as the sunlight is refracted by their ice crystals to produce an impressive spectacle of rainbow colours.

There are, in fact, two species of cirrostratus, which look different from each other. *Cirrostratus fibratus* is made up of delicate, parallel fibres and *cirrostratus nebulosus* is smooth-looking, like a veil, so there are no tonal variations to help you spot it.

Cirrostratus clouds are transparent and hang high in the sky like a lighter blue/milky whitish blanket.

This cloud signals the arrival of a warm front, usually followed, within 24 hours, by rain.

As they are so thin, you will always be able to see the sun through this cloud.

There are two varieties of cirrostratus: *undulatus* looks like waves and *duplicatus* means there is more than one layer at different altitudes.

Moody and sombre-looking

NIMBOSTRATUS CLOUDS

These are the clouds nobody wants to see. This cloud, hanging about like a wet blanket, promises rain, rain, rain, perhaps some sleet and maybe even snow.

These thick grey rain clouds give the skies a bad name. Firstly, they block out the sun – entirely – and then they drench everything with a steady downpour of rain that will go on for hours.

Of course, if your summer has been unseasonably hot and your gardens and crops and water supplies have suffered, you may have been rain dancing to bring in these clouds. If you've been praying for rain, then these clouds are a sign your prayers have been answered!

Although this is a low-hanging cloud, it forms in the middle atmosphere and then spreads upwards and downwards to dominate the skies. If you are prone to championing the underdog and the overlooked then this might have to be your favourite cloud.

The prefix 'nimbo' comes from the Latin word *nimbus*, which means 'cloud' or 'halo'.

It's not all bad news: they don't bring thunder and lightning. Just rain, rain, rain.

Nimbostratus is generally a sign of an approaching warm front, but this can take days to arrive.

These clouds are so featureless, there is no further subdivision into species or varieties.

Rare and fleeting

CIRROCUMULUS CLOUDS

The rarest of the ten main cloud types (page 16), these clouds are formed of tiny cloudlets that don't like to hang about, which is what makes them so hard to spot. Look for small round 'puffs' of white or sometimes grey cloudlets with a grainy texture and organised in rows or patches.

You can also try the finger-width test, which states that these high-level clouds should be no bigger than the width of your littlest finger when you stretch out your arm and point it upwards to the sky.

Cirrocumulus clouds are composed mainly of water droplets which, if cold enough, then freeze to form ice crystals – this is because they occur at higher temperatures but lower altitudes than cirrus (page 50) or cirrostratus clouds (page 54).

If you fly through cirrocumulus clouds, it will appear as if you are passing through a thin fog. Expect light turbulence. If you fly above them, the cloudlets below will have soft outlines resembling cotton-wool balls.

Some say the grainy texture and layered appearance of these clouds resembles a honeycomb. Others see more similarity to the gentle ripples of an ocean.

Cirrocumulus can also look a bit like fish scales, hence the name 'mackerel sky' when they appear.

These are high-level clouds that occur above around 3 km (10,000 ft) in polar regions, 5 km (16,500 ft) in temperate regions and 6 km (20,000 ft) in tropical regions.

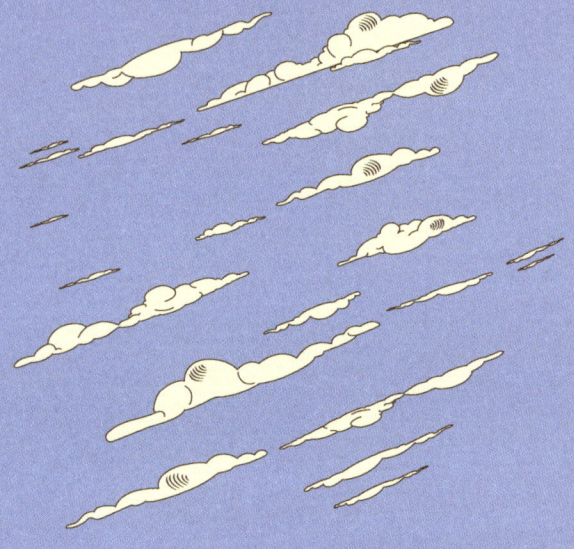

Scrappy, commonplace and blocks out the sun

STRATOCUMULUS CLOUDS

The sky is overcast and when you look up, you can see the bottom of these pesky sun-blocking clouds look grey and dense. Welcome to the cloudscape of the low-lying stratocumulus, which has the distinction of being the world's most common cloud.

There are three species – *stratiformis*, which extends over large areas of the sky; *lenticularis*, which can look like lens-shaped spaceships hovering; and *castellanus*, where the top layer of the cloud rises in turrets.

And there are even more varieties (seven, in fact) which include *opacus*, which will completely obscure the sun and the moon; *undulatus* which – as the name implies – is wave-like; and *lacunosus*, where you will see large holes fringed with clouds when you look up.

The good news about this degree of variety is you can let your imagination run wild. And if you are in an aeroplane and flying over stratocumulus, what you will see is a dreamy fairytale landscape full of imaginary possibilities.

These low-level clouds have clumpy, well-defined bases which are usually white or grey.

This cloud can extend across what looks like the whole sky, casting gloom and making the day feel like one for staying indoors.

On the plus side, stratocumulus clouds can give way to beautiful crepuscular rays – these are (usually) orange sunbeams that appear when the sun is just above or below a layer of clouds at dawn and dusk.

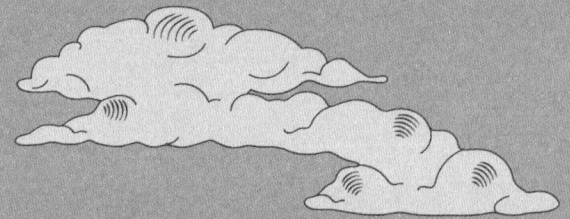

Shrouded in gloom

STRATUS CLOUDS

There's something a bit bleak about the stratus clouds, which are the lowest-hanging of all the cloud types. You know when the top of a building disappears behind a shroud of grey or when the thick fog lifts and all you can see is a grey and gloomy sky? This will be the featureless (unless misery is a feature) stratus clouds.

The word stratus is Latin for 'layer', and so these clouds are characterised by the appearance of horizontal layering with a uniform base. These dreary clouds, which can extend for many kilometres and last for days on end, can deliver drizzle or light snow.

Probably the best thing we can say about this cloud type is that it's atmospheric, and the perfect cloud cover if you're thinking of making a psychological thriller. For example, *The Lodger: A Story of the London Fog* is a 1927 Hitchcock silent film, which tells the tale of the hunt for a Jack-the-Ripper-type serial killer and uses the idea of thick stratus cloud cover as a sinister shroud to hide what's going on in the back streets of London every Tuesday night, when the barbarous killer is out and about.

Stratus clouds form in calm and stable conditions, when a gentle breeze lifts cool and moist air up and over colder land or ocean surfaces.

This is the only cloud that ever drops as low as ground level and when that happens, we call it fog.

These clouds appear as flat, hazy-looking layers and range in colour from dark grey to almost white.

Stratus clouds completely block the sun and so have a net cooling effect on the ground temperature.

Diverse and dramatic

ALTOCUMULUS CLOUDS

There are at least four species and some seven varieties of altocumulus, which tells us this mid-level cloud type can come in many different shapes and sizes.

Made up of cloudlets that hang in layers or patches, these clouds are usually grey or white and are distinguished from the cloudlets of cirrocumulus (page 58) because these are shaded on the side away from the sun. They are also much bigger, appearing between one and three fingers' width when you stretch out your arm and measure them with your hand.

The species *Altocumulus stratiformis* stretches over large areas of sky and does not appear in patches, while the species *Altocumulus castellanus* appears with a top layer formed in the shape of castle turrets.

These clouds can be thin enough for the outline of the sun or moon to show through (translucidus) or they can be thick enough to mask both (opacus).

Altocumulus clouds are made mostly of liquid water, but don't usually produce rain; rather, they can indicate fair weather.

In the rays of a low sun, altocumulus produce the dramatic, beautiful cloudscapes.

The base of altocumulus sits in the mid-level of the atmosphere, between 2–6 km (1.2–3.7 m).

In the summer, the appearance of the turret-topped Altocumulus castellanus may indicate change coming, with the weather turning from pleasant to severe.

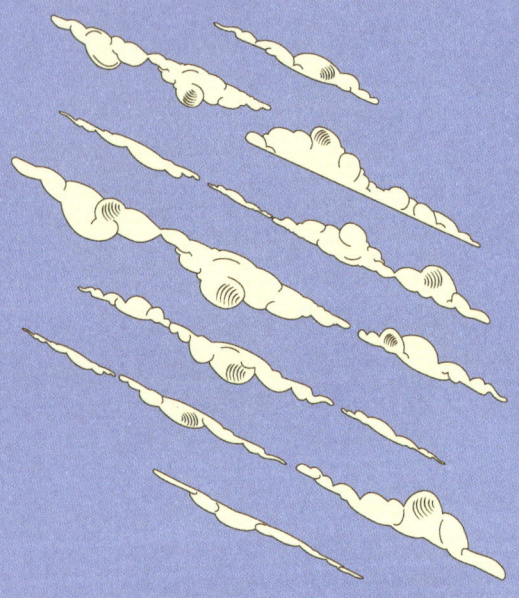

Overcast and dull as ditchwater

ALTOSTRATUS CLOUDS

Never judge a book by its cover or a cloud by its sorry-looking appearance, because while by day altostratus is probably the dullest of clouds, at sunrise and sunset it can be a whole different story.

For what may feel like the briefest of glorious moments, our normally dreary altostratus sky can light up with a spectacle of gold, red and purple and is so inspiring you will want to capture and keep a memory of it forever. Unfortunately, as the day dawns or ends, our cloud returns to . . . well, boring.

These darker-looking altostratus clouds form, most commonly, when a high-level cirrostratus (page 54) begins to thicken and are the result of a large region of warmer air pushing against one of colder air.

The name comes from two Latin words: *altum*, which means 'height', and *stratus*, which means 'flattened' or 'spread out'.

Altostratus is so forgettable it does not even have different species and even the varieties – 'Translucidus' (thin enough to show the sun and moon), 'Opacus' (thick enough to mask the two), 'Duplicatus' (more than one layer), 'Undulatus' (wave-like) and 'Radiatus' (rows that converge towards the horizon) struggle to make this cloud type a showstopper.

Best known for doing nothing much at all, altostratus clouds can sometimes produce light drizzle or even delicate snowfall.

Forget spectacular phenomena like halos or sun dogs, because these thin, mid-level grey clouds offer no tricks of the light to entertain us.

Both stratus (page 62) and altostratus clouds are thin layered clouds, but the latter forms at a higher altitude and so unlike stratus clouds, they don't touch the ground or shroud tall buildings in gloom.

Drumroll . . . the superstar of all the clouds

CUMULONIMBUS CLOUDS

If it's drama, impact and presence you're seeking from your cloudy skies then look no further than cumulonimbus – also known as thunderclouds – which is often described as the rock star of the main cloud types.

This is the tallest of all the clouds and it produces the heaviest showers, some of which throw the drama of hailstones in the mix, along with noisy thunderclaps and dramatic lightning shafts.

Look for a dense and towering low-lying cloud that can rapidly build itself upwards, thanks to the buoyancy of powerful air currents flowing in that upwards direction.

Pilots are forbidden from flying through cumulonimbus. This is because the updraughts that form inside a supercell can be the equivalent of a weak hurricane. The turbulence inside the cloud can be extreme enough to break an aircraft apart.

So, while these are the most exciting clouds in the sky, they are also – with good reason – the most feared!

From a distance, this looks like a calm white mushroom-shaped cloud that wouldn't say boo to a goose. But don't be fooled . . .

Cumulonimbus can join forces with its neighbours to form mighty 'supercell' storms.

Cumulonimbus clouds are most common in tropical skies.

A dust storm caused by a cumulonimbus downpour is known as a haboob.

Looks like alien spacecraft and hovers near
mountains and hills

LENTICULAR CLOUDS

These lens- or saucer-shaped clouds look for all the world like UAPs ('unidentified aerial phenomena' – what we once called UFOs) as they hang above our heads, beckoning our imaginations to take flight to strange and alien worlds beyond our own. But if you want to be less fanciful and romantic, you can visualise lenticular clouds as looking like a stack of breakfast pancakes.

Lenticular clouds form when high winds hit a tall upright structure – like a mountain or a human-made structure. This physical obstacle can divert the air up and over it and as this air rises, it cools. When there is enough moisture present, it condenses into these flat and otherworldly clouds.

On the sheltered side of the obstruction, the air moves up and down in a way that is similar to the wave-like ripples that form on the surface of a pond when you throw a stone into the water. As it rises it cools and condenses, and as it drops into the trough of the wave it turns to vapour and disappears. This means lenticular clouds are really an optical illusion. They appear to be stationary but in fact are moving, being formed and dissipating along with the undulating airflow.

Lenticular clouds are most common on high ground and in mountainous regions, but they can form in low-lying non-mountainous regions if the winds are strong enough.

Because of their curved shape and where they hang out, people often mistake these clouds for alien spacecraft.

These clouds can appear suddenly – and just as quickly disappear – which only adds to their marvellous mystique.

**Lumpy and bumpy pouches that usually signal
the storm has passed**

MAMMATUS (MAMMA) CLOUDS

Mamma clouds hold the distinction of not being a cloud type, genus, variety or species – rather they are what are known as 'supplementary' clouds, because they are always attached to other cloud types.

Aviators are warned not to fly through mamma clouds because they are most usually attached to the bottom of severe storm clouds including cumulonimbus thunderclouds (page 68). That said, they may also be found hanging off altocumulus (page 64), altostratus (page 66), cirrus (page 50), cirrocumulus (page 58) and even contrails (page 31). It is a misconception that they are a sign of a storm brewing; the fact is they usually appear once the worst of that storm has passed over.

In Latin, the original language used for cloud taxonomic classification, *mamma* means 'breast, udder or mother', and if you are lucky enough to see these clouds for yourself, you can't miss their mammary pouch or lobe-like appearance.

Mamma clouds were first described by a Bristol-born English clergyman and pioneer meteorologist, William Clement Ley (1840–1896), who studied the dynamics and use of clouds in weather forecasting.

Usually made of ice, mamma clouds can also be a mixture of ice and liquid water or made up almost entirely of liquid water.

Each lobe or pouch of the mamma cloud will be typically between 1.63.2–3.2 km (1–2 miles) across and visible for about 10 minutes.

Their distinct-looking lumpy undersides form when cold air sinks down and forms pockets, in contrast to those puffy clouds formed when warm air rises.

Rare, with an otherworldly beauty

NACREOUS CLOUDS

These showstopper clouds make their spectacular appearance in winter, forming some 16–30 km (10–20 miles) up and towards the poles, which is why they are sometimes known as polar stratospheric clouds.

Most common in places like Canada and Scandinavia, nacreous clouds are illuminated by sunlight from below. Their beauty is ethereal, with that kind of luminescence we associate with mother of pearl.

Made from tiny ice crystals which can bend (refract) the light, look for bright iridescent colours streaking through the lower clouds, as sunlight shines through nacreous formations at the start (after sunrise) and the end of the day (before sunset).

Nacreous clouds form when the atmosphere is so stable that the waves of air that form over mountains push moisture up into the stratosphere, but their beauty comes at a great cost because their ice crystals accelerate the destruction of the ozone layer caused by the chlorofluorocarbon (CFC) gases we pollute our environment with.

Nacreous clouds form in those higher and cooler layers of the stratosphere where the temperature is -85°C (-121°F).

The colours of nacreous clouds look like those you see when a thin layer of oil floats on top of a puddle of water.

The Latin word nacre means 'mother of pearl', so this is where they get their name.

CFC gases are produced by aerosol sprays, refrigerants, solvents and packing materials.

Castles in the sky

CASTELLANOS CLOUDS

'Castle in the sky' is an English idiom that suggests someone is chasing something highly unlikely to happen. But it turns out that castle shapes in the sky are not the stuff of fairytales or wishful thinking.

Castellanos clouds are not a cloud type in their own right but a species of the cloud types we've already explored.

They form tunnels or turrets on the top surface of the main body of the cloud and can appear with both the high- and low-level clouds such as cirrus (page 50), cirrocumulus (page 58) and stratocumulus (page 60), but it is in the mid-level of the atmosphere and the altocumulus (page 64) clouds that they are easier to spot.

There is a good chance that altocumulus clouds with these defined turret formations will continue to grow and may even go on to become stormy cumulonimbus clouds (page 68).

Some scientists now argue that castellanos clouds should not be a species but should be reclassified as a genus (main cloud type) in their own right.

Castellanos clouds can be harbingers of big showers and thunderstorms, making them a favourite among storm-chasers.

These crenelated clouds are always a sign the conditions in the mid-level atmosphere are unstable.

You can think of these shapes as castle turrets, with the cloud's protrusions being taller than they are wide.

Glider pilots who can use castellanos clouds to soar higher on thermals have nicknamed these clouds 'rocket clouds'.

77

Ocean-like waves or ripples in the sky

UNDULATUS CLOUDS

If you look up and see the surface of a cloud that looks just like waves or ripples, then you are looking at a cloud variety known as undulatus – also known as wave or billow clouds.

This variety is fairly common and can appear in six of the ten main cloud types, as follows: cirrostratus (page 54), cirrocumulus (page 58), stratocumulus (page 60), stratus (page 62), altocumulus (page 64) and altostratus (page 66).

Undulatus will form when the air both above and below a cloud type is moving either at different speeds and/or in different directions. Wave clouds also often form when mountains or islands force air to rise. As we know, air cools as it rises and produces clouds when there is enough water to condense. Once over the 'obstacle', the falling air temperature rises, but the initial disturbance creates a propagating wave pattern.

Rolling undulatus clouds can be puffy (cumulus) or sheet-like (stratus).

The waves will look uniform, and you may sometimes see a double system of undulations.

They form perpendicular to the direction of the wind. Don't confuse them with cloud streets (page 83), which run parallel to the wind direction.

They signal likely rainfall over the next 24 hours or result in an overcast day.

These can look like a fishing net in the sky

LACUNOSUS CLOUDS

Lacunosus clouds form when the cooler (denser) air above a cloud sinks down into 'pockets' through the warmer air of the cloud itself, creating a pattern that looks for all the world just like a honeycomb or a fishing net.

These are rare and short-lived 'clouds', which technically are not really clouds but the gaps you can see between puffs of actual cloud. You can think of them as being a cloud variety.

Lacunosus formations are most commonly linked to cirrocumulus (page 58), then altocumulus (page 64) and less often stratocumulus (page 60). With the high-altitude cirrocumulus, they will look like holes between the cloudlets, with the mid-altitude altocumulus you'll see them as gaps between cloud heaps and with the lower-altitude stratocumulus you'll spot the gaps between the puffy cloud layers.

These beautiful patterns form at all three altitude levels, but they will be hard to spot as they are so short-lived, lasting just a few minutes.

The word lacunosus comes from Latin and means 'full of gaps'.

The holes you see will be round with frayed edges; these are the edges of the actual cloud.

You may look up and think these clouds make the shape of a honeycomb pattern in the sky.

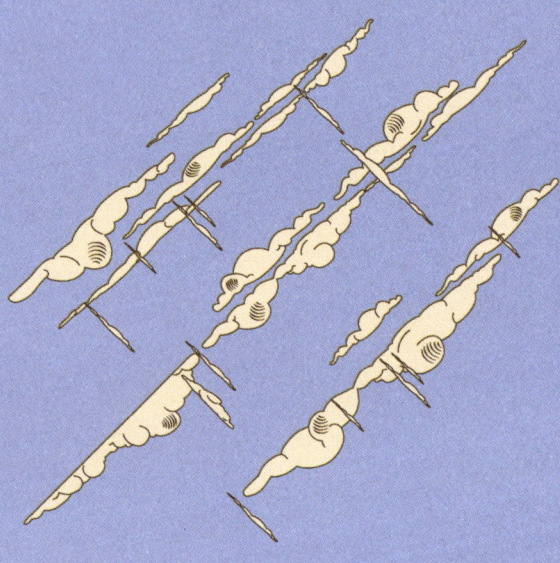

Clouds appear to radiate out from some distance

RADIATUS CLOUDS

Sometimes known as 'cloud streets', depending on which cloud type they fan out from, radiatus clouds are another cloud variety, rather than type or species in their own right. These parallel lines follow the direction of the wind; when they are perpendicular, they are the undulatus variety (page 78).

For those seeking real drama in the skies, the most impressive radiatus clouds are those in the higher altitudes that form jet streams – those 180 mph (290 kmph) winds that whip around the globe. Here, the cirrus clouds are formed of ice crystals (page 28) and the rays of the radiatus variation, which run in parallel bands and strips, appear to converge and disappear over a distant horizon hinting at another undiscovered world we will never reach.

If you have ever seen the UK Royal Air Force's Red Arrows fly by, you may have witnessed the radiatus effects of the contrails (page 31) as they too disappear over the beckoning horizon. Their red and blue vapour trails are created by injecting a mix of 75 per cent diesel and 25 per cent dye into the hot exhaust of the Hawk aircraft's jet engine. The white vapour is produced using diesel alone.

It is almost impossible to photograph the divergent lines of radiatus clouds because they cover such a vast expanse of the sky.

Glider pilots will think their birthday has come early when they spot radiatus clouds, because they know they will reach higher altitudes by riding the cloud streets.

Multilayered and hard to spot

DUPLICATUS CLOUDS

Duplicatus can appear with the following five of the main cloud varieties – cirrus (page 50), cirrostratus (page 54), stratocumulus (page 60) altocumulus (page 64) and altostratus (page 66). They are common, but identifying the duplicatus cloud variety is advanced cloud spotting because frequently, that second or third layer simply isn't visible to the naked eye.

When the duplicatus cloud is a fibratus (page 90) variety of cirrus or cirrostratus then you may be luckier and be able to see the layers. This is because the wispy tendrils of the different layers will be travelling in different directions, making it easier to see there is more than one layer of cloud present.

When the sun is very low in the sky, you will find it easier to identify the double layers of the clouds because the Earth will throw a darkening shadow on the lowest layer and the sun will lend gorgeous ruby hues to the upper duplicate layer(s), creating what we all enjoy as a stunning sunset.

Hoping for that spectacular sunset? Pray for the mid-level and less opaque altocumulus duplicatus clouds in the sky.

These will appear as cloud patches, sheets or layers at slightly different levels, sometimes merged.

If aliens are your thing, add lenticular (page 70) to your wishlist, because these clouds will look like spacecraft from another world stacked on top of each other.

**Look for a cap-style cloud sitting atop
another cloud**

PILEUS CLOUDS

Blink and you might miss one of these beautiful but short-lived accessory clouds (page 34), which appear when warm air rises over an obstacle that just happens to be another cloud, usually cumulus or cumulonimbus (pages 52 and 68).

Pileus clouds usually form when drier air with a higher condensation level prevents vertical growth and instead leads to the horizontal spread which gives these gorgeous, iridescent accessory cloud types their name.

The appearance of pileus clouds means the parent cloud is growing rapidly, is full of moisture and is highly unstable. So, while you may be blown away by the beauty of one of these fleeting clouds, the next thing you know you'll be being blown away by the bad weather – a pileus atop a cumulus cloud usually means the latter will soon develop into a cumulonimbus, which we know means nasty thunderstorms.

Pileus clouds will magically disappear after a few short minutes, when the cloud beneath them rises and absorbs them.

The word *pileus* is Latin for 'cap'.

Pileus clouds can also form above mountains and ash clouds above erupting volcanoes.

The shimmering beauty of the pileus cloud is formed when sunlight diffracts through its constituent water vapour.

**Look for a thin 'ribbon' wrapped
around a main cloud type**

VELUM CLOUDS

This cloud is another accessory cloud, meaning it doesn't show up solo but linked to one of the main cloud types, usually cumulus (page 52) or cumulonimbus (page 68). But unlike its cloud cousin – the beautiful pileus cloud – the velum is very dull to look at.

It appears as a thin, horizontal patch of cloud – usually grey or white – hanging around the middle or the sides of the parent cloud, so it can appear to be separate or mixed in with the main cloud. It will always look darker, which will make it easier to identify and unlike pileus, it can cover a wide expanse of cloudy sky.

Velum clouds form when the parent cloud meets a stable layer of air during its growth phase. This results in the condensed air spreading out into what looks like a veil. Velum clouds are not common and because of their appearance, can be easily confused with the unassuming and much more common main cloud type altostratus (page 66).

The name comes from the Latin word for 'veil', so that's your keyword with this cloud type.

You'll most likely spot a velum cloud just above or at the sides of a group of other bigger clouds.

When velum clouds appear with cumulonimbus, expect thunderstorms.

When velum clouds appear with lower-level cumulus clouds, the conditions will be stable and fair.

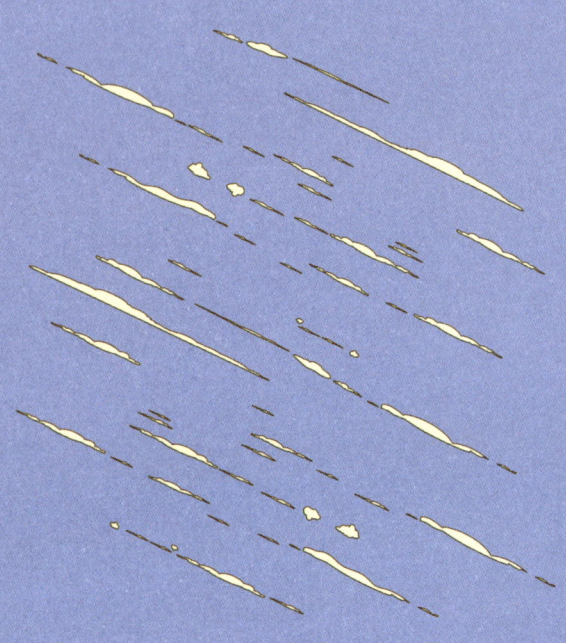

**Thin, delicate-looking and orderly strands
(or filaments) of high cloud**

FIBRATUS CLOUDS

These beautiful clouds appear when the ice crystals that make up the high-level cirrus (page 50) and cirrostratus (page 54) are pulled by the equally high and strong winds into long combed-out tendrils or streaks.

Fibratus is a species not a type or variety of cloud and doesn't really do much more than make the sky look pretty. It does not tell us anything other than there are strong and continuous winds in the higher altitude, through which the cirrus and cirrostratus clouds are dropping down.

With a cirrus cloud you will see thin, wispy-looking streaks and with the cirrostratus type, you will see more of a pale, veil-like layer of cloud. Cirrus fibratus clouds can also be formed by sheared aircraft contrails (page 31).

Fibratus clouds might mean there is a warm front on the horizon, followed by bad weather, or they can equally mean it's going to be nice out.

This cloud species is a favourite among cloudspotters, especially at sunset when it can give the skies an ethereal appearance.

Fibratus may bring out the phenomenon around both cirrus and cirrostratus clouds (pages **50** and **54**). This is because they are made from refracting ice crystals.

91

**Ominous looking clouds that
can look like smoke**

PANNUS CLOUDS

Also known as 'scud clouds', these are not likely to make you want to get your art supplies out. Rather, these are the miserable and menacing-looking accessory clouds (page 34) that make thunder clouds look even more alarming.

Pannus clouds sit at the base of rain-making clouds, especially cumulonimbus (page 68) and nimbostratus (page 56), but they can also hang about with cumulus (page 52) and mid-altitude altostratus (page 66). With the latter, you'll look up and see an uninspiring grey layer.

You can think of these clouds as being your 5-minute warning signal that it's about to tip down with rain, because although they don't produce precipitation themselves, they always come with clouds that do.

The pannus is a cloud accessory translated from the Latin word 'rag' and the Greek word 'shred', so look for a dark, ragged base to the clouds overhead.

These clouds may be attached to another cloud or sitting just below it.

Pannus clouds will frequently become more numerous and soon merge into one another to cover a bigger patch of a darkening sky.

If it's not already raining or snowing when you spot the sorry-looking grey of a pannus accessory cloud, it soon will be!

It will be raining

PRAECIPITATIO CLOUDS

Think of all the (natural) things that can fall from the sky and you'll start to understand what to expect when praecipitatio clouds show up: rain, freezing rain, ice pellets, snow, hail and ice crystals. You might want to wear a hard hat if you're out when these accessory clouds show up, or at least make sure you have your umbrella.

Precipitation forms when the suspended water particles that make up the main cloud type condense into larger droplets that are too heavy to be suspended any longer and so fall to the ground.

This cloud feature can appear with six of the ten main cloud types: altostratus (page 66), nimbostratus (page 56), cumulonimbus (page 68), cumulus (page 52), stratus (page 62), and stratocumulus (page 60) and the type of precipitation you will get depends on the temperatures of the various layers of air it falls through on its way from the main cloud to the ground.

This accessory cloud takes its name from the Latin meaning 'I fall'.

The precipitation must reach the ground to be considered praecipitatio. If it evaporates it's a different cloud feature known as virga (page 96).

Long ago, scientists believe it rained non-stop for a staggering 2 million years.

There are so many reasons to be grateful for rain – the main one being no rain, no life on Earth!

Look for rain clouds with wispy tails

VIRGA CLOUDS

If you're not hooked on cloudspotting by now, then seeing the virga supplementary clouds might just tip the scales – after which, there'll be no turning back.

These clouds are what are known as supplementary cloud features (page 35) and so what you are really looking at is a cloud that is producing rain or snow that never actually reaches the ground because it evaporates before then. This means then virga is linked to clouds that form at higher altitudes.

You will see a main clump of cloud with jellyfish-like trails of precipitation that hang down from the underside of the cloud. Virga can also lead to what is known as microbursts – where a localised column of air within a thunderstorm sinks downwards – which are more common but less well-known than tornadoes and which can be very dangerous to pilots and aircraft.

Virga clouds are more commonly seen in isolated areas like the desert, where they can appear beautiful and delicate; in other locations they are more likely to look grey in colour.

Virga clouds take their name from the Latin meaning 'rod' or 'branch'.

These clouds are at their most spectacular when seen at sunset and in conditions where a light wind extends and whips the tendrils into a pleasing curve shape.

Look for a menacing sometimes tatty-edged
cloud barrelling along at the front of a fierce
storm cloud

ARCUS CLOUDS

This is a low, horizontal cloud formation that will be linked to those big bad storm clouds, cumulonimbus (page 68). They form along the leading edge or gusty fronts of thunderstorms and, technically, come in two types – arcus rolls and arcus shelves.

The shelf version is wedge-shaped and attached to the base of the main cloud, the roll is much rarer, not attached to the main cloud and appears to be barrelling along like a solitary wave in the sky with a single crest. Neither is a good sign!

Arcus clouds form when a cold downdraft from a cumulonimbus cloud reaches the ground, at which point it will rapidly spread along the ground, pushing the warmer and more moist air upwards. As this air rises, water vapour condenses into the patterns associated with arcus clouds. The new cloud may roll – which looks spectacular – if it experiences different wind directions above and below.

This is a supplementary cloud formation that takes its name from the Latin word *arcus*, which means 'arch'.

The 'Morning Glory' cloud, which shows up each autumn in the Australian outback town of Burketown. It is a gigantic roll cloud that is hundreds of kilometres long.

These clouds form in the lower atmosphere at a height of up to 2 km (6,500 ft) and are some of the most photogenic of all the clouds.

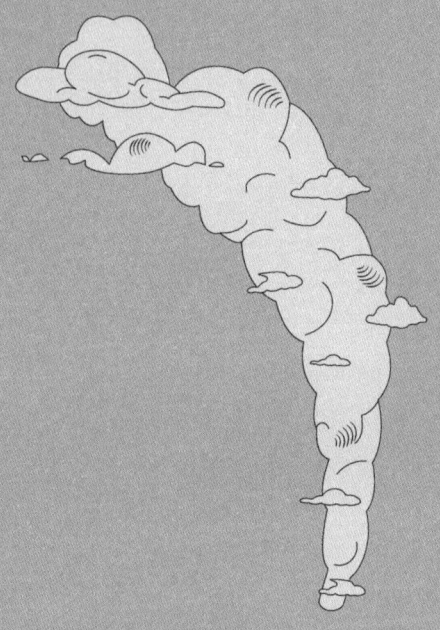

Looks like a finger pointing down from the sky

TUBA CLOUDS

Tube clouds (aka funnel clouds) are formed by the vertical stretching of something called vorticity. This is the name given to the air moving in a spin through the atmosphere – a phenomenon caused by wind shear. In simple terms, the latter is a change in either wind speed or direction over a short distance which can occur either vertically or horizontally.

Tubas form when air is sucked up from the ground into a cumulonimbus storm cloud (page 68) to 'feed' its rapid vertical growth. Cloudspotters describe this as being a bit like watching swirls down the plughole of the bath, only in the opposite direction because this rapidly rising air begins to spin and create a vortex. In a big storm, this rising air expands and then cools to form the walls of the tuba or funnel cloud.

While not all funnel clouds become tornadoes – some dissipate before hitting the ground and so don't cause the same kind of terror and damage – tornadoes can start their life as a tuba cloud.

The tuba or funnel cloud formation will look like a large finger projecting out and down from a menacing storm cloud.

Tubas can form when a single cumulonimbus storm cloud becomes a more coordinated system known as a multicell or supercell storm. This increased organisation creates the formation of the tuba.

Downward spiralling tubas where the air is sinking create dramatic but less harmful landspouts or waterspouts.

**Look for an anvil-shaped stormy
cumulonimbus cloud**

INCUS CLOUDS

This is another supplementary cloud feature and one that is always associated with the stormy cumulonimbus clouds (page 68). It is the formation at the top of the cloud that gives cumulonimbus its blacksmith's anvil shape and is, in fact, the single most distinctive feature of a storm cloud.

The science behind the formation of incus goes as follows: usually, air becomes colder the higher it goes, but when you see an incus formation, you know that the rising air that is forming a stormy cumulonimbus cloud has unexpectedly run into a 'ceiling' of warm air.

Suddenly, the usually warmed air inside the cloud is no longer a higher temperature than the air around it. It is also less dense and so can no longer rise up. Meteorologists call this a temperature inversion. This then forces the cloud to spread outwards – producing the incus (anvil) shape – rather than continue its rapid vertical growth.

These cloud formations come from the Latin word *incus*, meaning 'anvil'.

Incus clouds can cover hundreds of square miles of stormy-looking sky.

The temperature inversion that gives rise to incus clouds usually marks the boundary between the troposphere (page 15) and stratosphere (page 15). This is known as the tropopause.

If you spot an anvil cloud, expect heavy downpours, thunderstorms and even sometimes a tornado.

These clouds sit or wave over mountain tops

CAP AND BANNER CLOUDS

A cap cloud, as the name suggests, looks just like a hat sitting atop a mountain, hiding it from view. Described simply as an 'other cloud' type, it is really a form of the lenticularis clouds (page 70) but instead of forming downwind of the mountain, it forms over the top.

Cap clouds form in stable conditions as the air rises to scale the top of the mountain and cools down as it does so. Banner clouds – which form in the same place as cap clouds, but which can look more like a cloud flag fluttering in the wind – need a strong and turbulent wind forming behind a sharp summit to take their distinctive shape.

Both cap and banner clouds are described as 'orographic clouds'. This means they are shaped by the land below. The word comes from the Greek and means 'hill'.

Cap clouds usually appear stationary, while banner clouds will wave slowly in the wind.

A cap cloud is formed when humid air – moistened by the mountain terrain – is forced to flow over the mountain, condensing into a cloud.

Some mountains, such as the Matterhorn in the Swiss Alps are famous for their frequent banner cloud formations.

Look for a repeating breaking wave pattern

KELVIN-HELMHOLTZ CLOUDS

These breathtakingly beautiful clouds are super rare and, for cloudspotters, even more special for that. They can occur at all three cloud levels and are a version of the undulatus cloud variety (page 78) of three main cloud types: cirrus (page 50), altocumulus (page 64) and stratocumulus (page 60).

Catalogued as 'another cloud type', these clouds are named after the Kelvin-Helmholtz theory, which explores the instability inherent in a shear layer in fluid, which in the natural world gives rise to unusual phenomenon in both the oceans and the atmosphere.

These clouds form thanks to wind shear, when there is a sharp boundary between layers of colder air below and warmer air above, with the upper layer moving more quickly than the lower, creating the wave-like undulations that surfers dream of!

If you're lucky, you might spot these clouds as the top layer of a fog – but lucky is the word, because whichever cloud they are linked with, their appearance is always fleeting and will last only a minute or two.

These clouds can form at any height but are usually seen at higher levels.

Two scientists lend their name to this cloud formation thanks to their research into turbulent airflow. They are the physicists, Hermann von Helmholtz and William Thomson, Lord Kelvin.

Tiny ice crystals that glitter in the night sky

DIAMOND DUST CLOUDS

This phenomenon is the stuff of romantic dreams and fairytales. Made up of tiny ice crystals that are so small they appear to be suspended in the air above, these cloud formations twinkle and glitter like slowly falling fairy dust in the night sky.

Diamond dust can only form in extremely cold temperatures (less than-30°C/-20°F) so that the water vapour in that cold air stays as floating ice crystals rather than condensing into droplets which freeze. The two are not the same.

Because they need such low temperatures, they are most common in the polar regions where they do more than glitter and freeze, they also reflect and refract the light that passes through these ice crystals to produce a range of spectacular halo phenomena including magical arcs and halo rings.

Diamond dust is actually a beautiful precipitation falling through a very clear night sky.

These low-level clouds are sometimes called ice-crystal fog, although they don't obscure visibility in the same way as a true ice fog would.

These beautiful clouds fall slowly, like household dust, and reflect the light, making them sparkle just like diamonds.

Unlike ordinary snow, diamond dust can fall from a cloudless sky and on a sunny day.

With grateful thanks to publisher Kate Pollard, who had the original idea for this beautiful series of nature books, to illustrators Siena Zadro and Katherine Zhang and to our wonderful assistant editor – Phoebe Bath, who championed the clouds with an infectious enthusiasm and skilfully steered the words and illustrations from concept to conclusion with us. I cannot wait to tell the clouds there is a new book celebrating them in all their magnificence.

Susan E. Clark spent a childhood in the fields, hedgerows and streams of the North Devon countryside and has been a Nature-lover ever since. She studied biological sciences at University and is the editor of *Resurgence & Ecologist*; the flagship magazine for the environment which champions the natural world and the interconnectedness of all who share the planet.

Quadrille, Penguin Random House UK, One Embassy Gardens,
8 Viaduct Gardens, London SW11 7BW

Quadrille Publishing Limited is part of the Penguin Random House group
of companies whose addresses can be found at
global.penguinrandomhouse.com

Penguin
Random House
UK

Published by Quadrille in 2025

www.penguin.co.uk

A CIP catalogue record for this book is available
from the British Library

ISBN 978-178488-981-4
10 9 8 7 6 5 4 3 2 1

Publishing Director: Kate Pollard
Editor: Phoebe Bath
Design and Art Direction: Evi-O.Studio | Katherine Zhang
Illustration: Evi-O.Studio | Siena Zadro, Katherine Zhang
Production Controller: Martina Georgieva

Colour reproduction by p2d

Printed in China by RR Donnelley Asia Printing Solution Limited

The authorised representative in the EEA is Penguin Random House
Ireland, Morrison Chambers, 32 Nassau Street, Dublin D02 YH68.